Turtle Poems

Turtle Poems

Revised Edition

Michael Christensen

Turtle Poems Publishing

Turtle Poems by Michael Christensen first appeared in print in 2004.

Some of these poems have appeared in the following publications: The Bridge; Behind BAARS (Journal of the Bay Area Amphibian and Reptile Society); British Chelonia Group Newsletter; Chelonian Conservation and Biology (Chelonian Research Foundation); Herptofauna (The Journal of the Long Island Herpetological Society); The League of Florida Herpetological Societies Newsletter; NCHerps (North Carolina Herpetological Society Newsletter); Pacific Northwest Herpetological Society Newsletter; Red Owl; Terrapin Tales (Mid-Atlantic Turtle and Tortoise Society Newsletter); Tortuga Gazette (California Turtle and Tortoise Club Newsletter); Tucumcari Literary Review; and The Turtle Tribute.

Contact: turtlepoems@gmail.com
More information: www.lulu.com/turtlepoems

The turtle prints featured in Turtle Poems are from "Tortoises, Terrapins, and Turtles: Drawn from Life," by James De Carle Sowerby, F.L.S. and Edward Lear, originally published in 1872. The complete volume is currently published, reprinted and sold by the Society for the Study of Amphibians and Reptiles. For more information go to: http://ssarherps.org/publications/books-pamphlets/facsimile-reprints

ISBN 978-1-304-68803-3

"Gracie was a turtle dog..." p. 63

To Alice and Aristotle,
one is my wife,
the other a turtle

Contents

Turtle Poems

[PLATE LI]

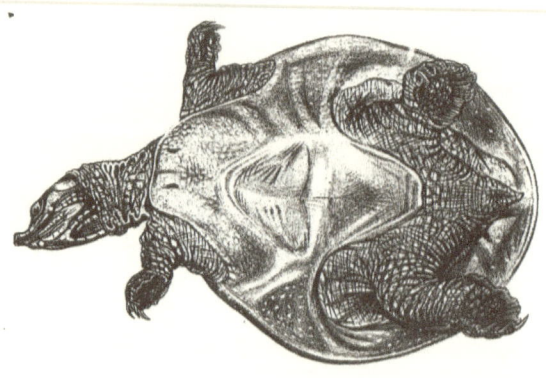

Ganges softshell turtle (*Nilssonia gangetica*)

Meat Market

Strange softshell turtles
large and flaccid,
stacked in plastic boxes
next to the sidewalk,
on a cold hard butcher shop floor.

Maybe you were
raised on a farm,
where captive born turtles
know nothing of freedom,
fattened for the kill –
your every need supplied.

More likely though
you were trapped this summer,
wrenched from your native watery,
the waters you will only know
in your last turtle dreams.

We are not your friends,
our actions do not deceive you.
Even I, empathetic to your plight,
will never be trusted.
How frankly you stare,
how brave you seem
in spite of these odds.

How I wished for a moment
I were as brave,
how I wished since that day
I could forget those eyes.

Pancake Turtle

The softshell turtle
is not soft by nature,
with razor sharp jaws
and flesh rending claws.
She'd look like a pancake
but for her ornery maws.

How very strange to see
pancakes swimming in streams.
Wouldn't help to put syrup
or even whipping cream,
to sweeten the nature
of something this mean.

Though I shouldn't say
that this turtle is mean,
it's all about self-defense.
From folks much smarter
in the turtle world, to their
descriptions, I humbly defer.
A tubular snout, a clown-like face,
a skin covered flattened shell.

More like a fish than a turtle,
but more like a turtle when still.
But, never miserable
when in the sand
buried, invisible,
but a pancake still,
not cooked very well.

Mixed Blessings

As one of those
who has been blessed,
with acquaintance to *Sternotherus*,
I was first aware of its *odoratus*.

The musk turtle
is one of those,
who leaves an imprint via your nose.

But once you get
to know them well,
they lose the need to give off smell.

To know them well
I have been blessed.
I'll speak fondly of *Sternotherus*,
though to know me well,
they could probably care less.

Musk Turtle Party

I know whom not
to invite,
or trust,
at the occurrence of
a musk turtle bust,
keep the alligator snapper
a safe distance from the door.

Portions of musk turtles,
and other turtles too,
were all found in
the snapper's excrement stew.

One thing you shouldn't do,
you shouldn't let him through.
He'll eat both host and hostess,
and if small enough, he'll eat you too!

Musk turtles know
to avoid,
or mistrust,
the appearance of
this certain large mouth with
just a hint of musk?
When the alligator snapper breaks cover,
the musk's party is over.

Why have the turtles fled?
It's the snapper's smell they dread.
Where snappers come around,
musk turtles are rarely found.

Their appetite should not be
personally taken,
their choice of meals are mere chance,
no small creature's forsaken.

So remember
this social
season,
in ponds, lakes,
or rivers, slow, wide
and lonely, musk turtle's parties from now on
are invitation only.

Musk Turtle Pond

Stinking Jenny and old Stinkjim
have a pond named just for them.
Musk Turtle Pond is its name, although
the musk turtles there are seldom seen.

Stinking Jenny and old Stinkjim
rose to the warm surface one night,
where branches dipped in the water deep,
amidst the din of frogs galore,
and scores of insects above the roar.

Said Jim to Jenny, with his eyes
have you ever felt such a night?
Said Jenny, *no*, so up they would climb,
onto a branch safe from the shore,
a young couple's boat seemed destined for.

The boat appeared out of the dark
the lantern on the bow gone dim.
Two fishing poles without bait, ignored
by two captives of a starry night,
pursuers of a summer in flight.

In the boat where the couple lay,
gazing out at the Milky Way,
suddenly branches were all around,
the total darkness where they were found.

All at once there came a crashing
as the turtles fell from their branch.
The couple thought of snakes in the trees
and teeth of creatures they could not see.

They wondered what had brought this stink
smelled by the humans on the brink,
of romance replaced by horror when,
he reached for the flashlight but found Jenny
and Jim.

While those two turtles' anger grew
toward the man who wished that he knew
what on his finger, sorely attached,
or to his thumb with a grip to match!

He shook his hand in disbelief,
above the water, dark and deep,
as she finds the flashlight and turns it on,
just two plops and the turtles were gone.

He received no more than a pinch,
and both the branches and skunk scent went.
He pushed their boat from the tangled shore,
him, never as bewildered before,
her, feeling she'd never loved him more,
despite this meeting down by the shore,
on Musk Turtle Pond down by the shore.

Dear Ernst and Barbour

I was wondering why
you identify,
common musk turtles
Sternotherus odoratus as
Kinosternon odoratum?

I'm sure you're very busy,
but if a reply could be arranged
as to why herpetological lit
published since,
does not recognize the change?

Since I went to the trouble
of finding your address,
I'm sure you wouldn't mind
if I digress. You see,
I know something of the social fringes,
where one separates
Sternotherus from *Kinosternon*,
on the basis of shell hinges.

It's when I lift my nose from your text,
and wish to discuss the change,
based on morphology, karyotype and
biochemistry it's then,
people find me strange.

But dear Ernst and Barbour
I don't find it strange at all,
I love the science of herpetology.
Whether it's crocodile, alligator,
turtle, tortoise, snake or frog,
lizard, salamander, newt,
or toad, come to crawl.

I don't believe the musk turtle,
or any of its herp-relations,
much care what we name them
it's all just human din.
Or how they're described
in reptile papers or books,
when paper meets the printer,
pencil or pen.

The name tells me not
to change this little turtle,
or even get in its way.
Because when the musk turtle must,
he'll give you a stinky dose of musk,
hence the name, *odoratus*.

Or *odoratum*, if science prefers.

With my digression now complete,
I'll look forward to your reply,
with respect and regards, goodbye.

To: Michael Christensen
From: Carl H. Ernst
Subject: *Sternotherus* vs *Kinosternon*
Date: 4/06/94

George Mason University

Dear Michael,
    The use of *Kinosternon* as the genera for the four musk turtles was premature. I have now switched back to *Sternotherus* (as have almost everyone else). My reasons are stated in the enclosed pages which are from the new edition (2nd) of *Turtles of the United States* which should appear about 1 September.

Best Wishes,
Carl H. Ernst

[Note: Carl H. Ernst and Roger W. Barbour have written many essential books and papers about turtles and tortoises.]

Goings on in a Pond

Musk turtle listening
to the snail's discussing,
what it is they've been
eating on the bottom.

Musk turtle waiting for
the snail's departing,
snail gets eaten before
his beg your pardon.

Flattened Musk Turtle

Sternotherus depressus,
flattened by nature,
I wonder who cares about
this small unassuming drab shell.
No heartstrings plucked by
its icy stare.

Sometimes it feels that people
of the world would not blink,
when told the startling news:
the flattened musk turtle is extinct.

And if they'd met you,
stream-bottom-dweller,
maybe they would've seen
that you were special,
and the places you once lived
were worth more
than a measure of coal.

Luckily not quite yet extinct,
there's still a chance
for us to meet.

But there are those given power
of deciding which animal will die,
to hasten the high-rise's rise.

Wealthy bankers are doling
out hard cash to the powerful,
funding final decisions of destruction,
leaving their offices by five.

Sternotherus depressus,
collateral damage
of financial gain,
measured now in amounts
of acid rain.

Hands still wringing,
cash registers still dinging,
and everyone's soul as empty as before,
if not a little more.

Mistakes we keep repeating,
digging more poisons from the ground.
Desiring coal's intense heat,
burning dirt for the demand of light to meet.

And above our heads
the sunset is red,
fog of coal smoke instead of fresh air
that we can breathe.

In those places, where coal is found,
keep it in the hallowed ground.
The world is better served in fact,
take the mined coal and put it back.

Sternotherus depressus,
if by extinction you left us,
how sad I would be.
Would the person
who was to blame
be held accountable?
Would they even know?

Musk Turtle Traits

You'll hardly ever see
musk turtles on a log,
with the common sliders
piled so precariously high.

But occasionally
you'll see one in a tree,
basking there half asleep
as if a bird it wished to be.

Then if you catch its eye,
from the tree it will fly,
safely above the water below
just a splash and off it goes.

If you're of a type
who has a beer too many,
you may rub your eyes thinking,
I should probably stop drinking.

Musk Turtle Vision

This aquatic turtle
is best suited when,
mesmerized by the mix
of its watery elements.
Musk turtle paradise.

I'm told musk turtles see
colors, science agrees.
I wonder if this trait
puts more turtle food on their plate.

Should you chance to meet the eye
of a musk swimming by,
or meet one high and dry,
their response will still be
they just don't care to see
colorful types like you or me.

Stinkjim

The musk turtle can be
both pugnacious and rude,
sometimes called a stinkjim,
for the scent it can exude.

Stinky little fellows
when frightened or captured,
and glad to give a nip
to an offered finger tip.

But that's not the first line
of defense in this turtle's mind,
first is to escape or hide,
never to be bothered.

The musk turtle can be
both elusive and shy,
and can stay in its shell
until the sun leaves the sky.

Interesting creatures
when kept inside a tank,
the will to stink goes away
but the pugnacious stays.

[PLATE LXI]

Red-eared slider (*Trachemys scripta elegans*)

Mister Red-Ear

Mister Red-ear
turtle green,
swimming in
a crystal stream.

He may see a
fish swim by,
have some lunch:
a drowning fly.

Mister Red-ear
eyes a frog,
has a bask
on an old log.

That's what Mister
Red-ear does,
when the sun
shines from above.

Now the day's done
time to sleep.
Turtles rest
in water deep.

The Trouble With Red-Ears

Then there were seven
then eight, then nine,
there's no end,
no fault of their own.
How can I abandon
these homeless red-ears
left to die?
Someone's pet,
that lived too long.

I just want to find them
a good home ...
they make great pets ...
we just don't have the time ...
we have to move ...
my landlord ... my husband ... my child's
no longer interested, no time....
The path to turtle rescue is
paved with good intentions.

Much maligned,
illegal in Florida,
bred in the south,
shipped by the thousands
around the world,
for pets and soup.

With fierce determination
the red-eared slider is here to stay,
established throughout the world
and multiplying.
Eventually there will be
nothing left
in the world's wetlands,
but mosquitoes, bullfrogs and red-ears.
This, to my own horror,
I gloomily predict.

Not these though,
not these nonnatives.
They will not be set free
to contribute to my
dire prediction.
Nor shall they ever return
to their native watery.

I will bring them home,
we will live out our days together.
I will commit my life,
as I did to my poor wife.

Red-Ear Christmas Turtle

The long winter nap
has been established,
for last year's beloved
Christmas gift turtle.

Elfie, aptly named,
green as a Christmas tree,
with jolly red swags
behind each bright eye.

As the days grow short
Elfie (who resembles an elf no more)
somehow knows,
that the leaves that settle
on the pond bottom
will be her blanket
from the cold.

Hibernation is determination,
a means to survive,
a profound acceptance
that some things will not change.

To awaken is the reward,
another compelling spring
to attract you from sleep,
and the blessed summer heat
to savor, always follows.

So Merry Christmas
sleeping turtle,
wish you were here but
we'll see you in the spring,
when you awaken
and part the clouds of winter.

Through Mud

Pleistocene eyes
of the muddy mud turtle,
you know today
not a thought to history.

If we could see
what the mud turtle has seen,
then maybe we
could finally see the trees.

Timeless, ageless
excelsior mud turtle,
rise from the mud.
Make the message crystal clear,

the past collides
with a day we called today,
when we awoke
and we knew before learning.

Pet Mud Turtle

In a fish tank where
a mud turtle calls,
the bubbling clear water
his home,

the fish can relax
and die of old age,
though the hungry mud turtle's
still found.

The fish needn't fear
just keep to his rear,
and he's forgotten that they
are there.

It makes no difference
nor injures his pride,
to be considered such things
as a rock.

What seems to matter
to this turtle's mind,
what hapless insect drowned or
fish croaked.

An Aberrant Mud Turtle

An aberrant mud turtle,
found in a place where
it isn't known to occur.

An aberrant mud turtle or
an aberrant world?
Rain, where it's normally dry.
Dry, where it normally rains.

An aberrant mud turtle,
waiting to cross a highway
that used to be a pond.
Kidnapped and released
a few states away.

An aberrant mud turtle
found in a place where
it isn't known to occur.
An aberrant mud turtle
left to endure.

The Natural

Alligator snapping turtle
a born fisherman,
on its tongue
its own tempting bait.

It just opens its mouth wide
in the mud where it hides,
and wiggles this pink appendage
like a lure,

so when a little fishy
believes it's found a bug,
it's down that mighty gullet
with one great gulping glug.

The Myth

So the myth goes
of the alligator snapper,
that with one powerful bite
can leave a wooden broomstick
one side to its left and
one side to its right.

Just a myth I suppose,
still you'll see no wiggling toes,
if the toes know,
where the alligator
snapping turtle goes.

Pond Bottom Slime

I wonder what's on the mind
of the old snapping turtle
down in the pond bottom slime
looking about carefully,
slowly spending its time.

What's probably on the mind
of the old snapping turtle
is the pond bottom slime is just fine,
so leave me to mine,
slowly spending my time.

Finally it moves, a dark shadow,
then just the tip of its nose
exposed above his star-like eyes
watching, driven to survive
and thrive in places we'd never venture.

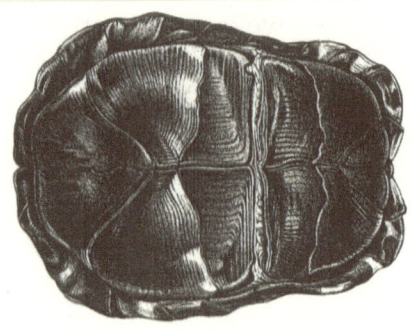

Eastern box turtle (*Terrapene carolina carolina*)

Box Turtle and Snail

Behold the box turtle.
Behold too the snail.
They both carry their homes
on their backs.

When these two fine creatures
leave this cruel world behind,
they'll leave two such quaint
little shacks.

But when the box turtle
is searching for a bite,
the snail must be quick
to conceal,

because if the box turtle
finds a snail at sunrise,
it won't be the turtle's
last meal.

Roxy

I know a turtle
her name is Roxy.
When she gets startled
she gets all boxy.

A hiss and then
she closes tight,
her head and legs
nowhere in sight.

Is it humans
you do not trust?
For all they've done
your fears are just.

And nighttime critters
will test you too,
I guess this world's
a scary place you.

At least you can
close up tight,
burrow in deep and
sleep well at night.

Close yourself like
a clam in its shell,
to be a box
has served you well.

Dawn

Comes the dawn.
Box turtle's eyes open,
they are red,
like the poisoned
skyline.

The turtle looks
then yawns,
absorbs the warmth
of the sun,
then responds
to a quickened pulse.

Hope springs eternal
that today will be
like the day before,
but the bulldozer
hastens grand plans
to pave paradise.
Tomorrow may not be
the same as yesterday.

Now the air cools,
driving the box turtle
back into its burrow.
It was a good day,
the bulldozer
did not come.

Beneath the Same Sun

Baby Bell's
hinge-back tortoise,
when just a few hours old,
you must unfold, from
your hunched position
in the egg.

From your first
tentative step,
to your first tiny bite, your
bright eyes see a new
world unfolding
before you.

From past times
when your ancestors
traveled beneath an African sun,
you recall something
about a vast
savanna.

To this time
so far from home,
a new life unfolds
on a continent unknown,
you search for a place
but an imprint is
all that's left.

Baby Bell's
hinge-back tortoise,
oh so tiny your steps,
you pause to smell
the earth beneath you,
then move on.

Painted Turtles in the Sky

Painted turtles in the sky,
where they must go
when they die.
No tear will fall
from sibling's eyes,
born alone, alone they die.

Those painted turtles in the sky,
painted turtles in the sky ...

Painted turtles on a rock,
when we appear
the turtles drop,
no time to say
or tell our thoughts,
that we intend to harm them not.

Those painted turtles in the sky,
painted turtles in the sky ...

Biting the Hand

When western painted turtles
are sunning on a log,
you'd probably think them
quite shy.

Falling into the lake,
at the slightest foot quake,
or when a shadow
of a creature
goes by.

If you take this turtle in
to your house for a swim,
and feed it a fishy
or two,

the shyness will be gone,
until what you have left,
a domesticated
turtle who,

you may think will take flight,
when it's offered a bite
from the hand that feeds it
above.

Best beware because turtle
cannot tell a finger
from a fish,
both likely seem

a palatable dish,
to the painted turtle
in your house for a swim,
that is,
if you take one in.

Green Turtle

Honu the great mariner,
appearing from
the fathomless depths,
rising on a current,
grazing on limu,
you eat your fill
then dream.

Turtle dreams,
floating away like
wood adrift on a sleepless ocean.
Once a tree, now lost at sea
unnoticed by turtles
swimming beneath
powerfully, seemingly
without effort. They glide
over the depths one can
only assume sometime ends.
The sandy bottom
a home for bones
of fallen ancestors to us all.

Turtle eyes
curiously oblivious to
the shore creatures.
Humans, with mask,
snorkels and ungainly fins,
amazed by your presence,
made happy by your
benevolent grin,
and hoping to stay,
when the surface called.

Happy turtle memories,
floating through my own dreams
of green sea turtles,
sending fish darting,
sorry to awake.

A Hole To Escape To

A gopher tortoise
won't stray far
from the safety of
its burrow.

It retreats with
exceptional speed,
at the slightest hint
of danger.

When some humans
speak their mind,
foolish things
begin spilling out like:

No global warming ...
No danger in fracking ...
Coal ash is safe ...
The Endangered Species Act
is bad for business...

Dangerous folk,
keeping a wary eye
on burrows beneath
mounds of cash,
ready to lumber
safely inside.

A hole to escape to
unto themselves
at exceptional speed.

The gopher tortoise,
a cavernicolous-
creature, sharing its
burrow home
with those who cannot
speak out in defense,
insisting retreat instead,
obligates to silence.

Real Property

The tortoise has no interest
in turning around or backtracking.
You can see it in the grim face,
that once a direction is chosen
there will be no wavering.

Determined even if
the choice is not wise.
I'll make it a wise choice,
the tort stubbornly says.

The tortoise doesn't care
to talk further of bad decisions.
We must remove that in which
a decision like this is made.

The tortoise does not care
to submit to the laws of dominion
or right of use, control, and disposition.
No borders in its mind.
Creating borders seems unnecessary.
I live on the land, says the tortoise,
I would never possess it.

It does not dream of possessions,
but seeks satisfaction,
and a leafy green.

If a tortoise dreams
it is not of regret or
bad decisions.
Dug in deep in its burrow,
dozing, it accepts the sound
and scurry,
of its nocturnal neighbors.

Turtle Communication

On the Orinoco
and Amazon river systems,
a word spoken,
so low in tone,
it requires a shell
to hear it.
Human ears not sensitive enough
to glean it.

A sound, a voice
there is a message,
transferring from somewhere deep,
onto the shell of another.

Turtles talking,
South American giant river turtles,
caring for their young
when it was thought
they ended their care
at the completion of the nest.

A mother turtle senses the state
of her nest, dug deeply in the sand.
She can hear the first babies calling out,
then replies in reassuring tones
that attract the babies, and encourages the
hatch.
The mother swims in the river below the nest
calling to them, she says,
It is time to come out, come out, come out,
it is time to follow, follow, follow.

The babies reply in high-pitched squeaks,
the babies hear and begin their first struggle,
even if they are not ready, they respond.

When the babies complete their hatch
they make their way to the water below.
Upon entering the water,
the babies move toward the mothers,
covering them and resting in the cavities
behind their legs and neck
and all over the carapace.

Many will hatch and follow the adults
to the feeding grounds,
flooded swamps, lagoons, and forests.

The food will be plenty.
They too may become something else's food,
this is how it has always been.

Our ears not sensitive enough to hear,
in captivity they go silent.

Just a Little Farther

At my work station
feeling a sense of freedom
from gazing upon
a picture of a wild turtle.

I saw the same turtle
for the first time
in a tropical fish store
in California.
They didn't know where
the turtle once swam free.

It can't be returned to
the freedom it once knew,
the freedom most of the world
only dream of knowing,
to a freedom I meant to give.

So I pay them the listed price,
will it buy the turtle some freedom?
Will it buy some for me?

I give it its own fish tank
cleaner and larger than at the store,
but I know in my heart,
it's not big enough.

I give it an even bigger tank
new filters and UV lights,
but I know in my heart,
it's not big enough.

I build it a pond in my backyard
where it can know the seasons,
though cannot venture
beyond the borders I've acquired,
but I know in my heart,
it's not big enough.
Just a little farther,
but I arrive somewhere
short of freedom.

Once on a walk,
I saw a turtle fall from a branch and
disappear deeply into the lake,
and I don't wonder
if it's free, or if the lake
is big enough.
Only glad for wild turtles.

Halfway Gone

Achilles shall never beat the tortoise
in a foot race,
so says Zeno's paradox.

Halfway, halfway, halfway gone,
through it all, tortoise lumbers on.

As each half gets smaller
it's all Achilles can do,
to keep up with the tortoise
who is doing nothing new.

Halfway, halfway, halfway gone,
through it all, tortoise lumbers on.

Will the developers beat the tortoise
in their race to
modernize the view?

Halfway, halfway, halfway gone,
through it all, tortoise lumbers on.

As the wild places shrink
to half of what we knew,
they want to halve them again,
what's half of nothing new?

Halfway, halfway, halfway gone,
through it all, tortoise lumbers on.

As each half gets smaller,
what more can the tortoise do?
If tortoise loses this race,
our fates will follow true.

Halfway, halfway, halfway gone,
through it all, tortoise tries to lumber on.

Turtle Heads

There's a pounding
in the back, out of view
but I know the sound.
A cracking of shell,
there is only one way
to get to the head,
you must injure it,
before you can kill it.

Standing a distance
from my pond,
a group of turtle heads
poke from just beneath
the green pond water.
It's that time of day.

The head will stay alive
after severed from the neck.
Gasping, biting, slowly
drifting into the Great Next.
It does not know
for up to an hour
that it is dead.
Its body has been quartered,
she will make a fine soup.

As I move toward the ponds
the turtles move toward me.
They all seem to know
that it is time to eat.
They do not know
that they will never be soup,
or that even such a thing
ever occurred.

Voice of the Turtle

When the last ice and snow
is melted by the rain,
and a humid silence settles
over the dead cattails,

when the first rays of sun
slice through the clouds,
and a familiar warmth flows
from the south,

something stirs in the mud,
something is unsettling the silence.

When the first frog wakes
and releases a hoarse call,
when a sleepy turtle rises
still covered in mud,

then an insect is heard
buzzing through the stillness,
and the turtle turns its eye
to the sound.

From the turtle's first yawn
to every waking creature's stirring,
the voice of the turtle
is heard in our land.

Turtles and Childhood

Sometimes two different words
seem to mean the same thing.
Like turtles and childhood,
summer and cotton sheets.

To some the turtle
holds the world on its back,
to others a pet
whose death is certain.
Is a child's neglect
that precedes a turtle's death
ever really forgotten?

For some a turtle
will live a long time,
from others, not a thought
can be spared
to turtles that existed
before humans could behold
this steady reptile stare.

If a turtle holds
this old world on its back,
its legs must be so weary.
When children say
goodbye to their youth,
sometimes a turtle is buried.

Myrtle, the Lost Turtle

I've lost my turtle.
Her name is Myrtle.
To her name
she doesn't reply.
Best rely,
on a steady eye.

Oh where, oh where
is that turtle?
I feel silly
calling its name,
so I silently search,
head down, as in shame.

Oh shame, oh shame
the bucket on its side,
the fault is mine
now I must waste time,
twisting my spine.

I lost my turtle,
and never did find Myrtle.
Looked all around
no sign of Myrtle.
Now simply referred to as
the lost turtle.

A Truth

And so a truth
shall now be known,
where cast a turtle
cast a stone.

The turtle gazes with
ageless eyes, clear and lucid,
fierce and ready,
knowing the past and
facing the future unblinking.
Suspended by the sun's warming rays.

As I try to get a closer look,
turtle suddenly sees me
and tumbles from
a log that fell
long before his sharp splash.

When startled turtles
tumble off fallen logs,
it's a reaction that
somehow reveals a truth.
The relationship between
humans and animals.

I wait until you surface,
your eyes ever watchful
for troubling movement.
The truth is, the world's
a dangerous place.
Today you survived
but the day is still young.

Tanks of Turtles

Lid lifting, suspense building,
turtles anticipating food.
A snail drops into a tank
of turtles hungry for snails.

The turtle's attention diverts
as if just thinking of something,
then looks and moves toward
the snail that is floating by.

Floating snail reaching for something
not water, the turtle trying to get his beak
around the smooth shell, instead
pushes the snail away, but follows.

The snail's shell snaps
like a nut between a cracker,
the turtle reminded what
the snail's been hiding.

The snail tries to escape
and not for one second
does it feel sorry for itself.

Once holes in garden leaves
cursing the snails to blame,
now a meal for turtles
and cloudy water is all that remains.

A strange assimilation,
where turtles from Florida
eat snails from France,
in California.

Marico Tortue

Unfamiliar sounds disturb my sleeping,
from the yard, into my window seeping.
Jumping from the bed and toward the door,
toward the sound feet hardly touching the
floor,

then stepping out the back door amazed, I
found
a group of wet creatures gathered round!
Worse, amidst this chaotic morning clatter,
my pond of turtles had all been scattered.

I chased the fur running every which way,
in the dim morning light I tried to survey,
when against the gate I follow a sound,
a knocking and a small turtle is found.

While sitting tightly in his dull gray shell,
a mud turtle chose to hide where he fell.
And to my dismay a large red-ear
her front leg chewed, blood everywhere.

With the final tally fresh in my mind,
there was still one turtle I could not find.
Was his fate within the creature's jaws,
or did he escape for some greater cause?
At the storm drain he did not even pause!

Then later on my bus feeling quite bad,
I thought to tell someone why I felt sad.
I've lost my turtle, paused behind my teeth
but from having said it in my mind,
you cannot say it, I realized
and not be a child of nine!
Just try it sometime.

But now, home from the vet thinking
about the fate of the lost turtle, hope sinking.
Around the pond I'm building a fence,
the only response that makes any sense.

I'm very sorry life turns out this way,
seems like things you love refuse to stay.
The title of this poem, the lost turtle's name,
I hope in escape, there was no pain.
Anyone could do the same *if Marico were me*,
you'd say, *I'd want to be free!*
Now where can that turtle be?

Feeder Fish

I must mention the fish,
never easy to use as food.
Some are colorful, and some are unique,
randomly, some set free each week.

They call them feeder fish,
they look like goldfish to me.
Somehow a goldfish is a step above,
feeder fish, never bred to love.

Out in my backyard ponds
a transformation's made.
The feeder fish have figured out,
keep away from turtle's snout!

So now they swim and grow,
mere feeder fish no more.
Shimmering gold beneath sun and snow,
and the turtles somehow know.

Food Chain

I placed twelve
rustling crickets
into a container.
They were food
for some box turtles.

When I returned
two days later
to feed the turtles
there was only one
cricket left.

One very large cricket
with no visible means
of escape for
the other eleven.

This one
very large cricket
may not have eaten
all the others

but it did dine
on the second to
the last
of that muted choir.

Then, with one or two
chews, a box turtle
is through
with all twelve.

Collections

Because I have turtles
my friends collect
strange creatures too.

Friends bearing gifts,
garden or water snails, slugs,
worms, pill and sow bugs.
Sometimes even spiders, in
reused containers they arrive.

Lifting the lid sending
spiders scurrying,
pill bugs rolling and
slugs to blinking.
I can't help but laugh.

I'm always amazed when
these gifts arrive.
How it was me
they thought of
while catching
these strange beasts
in their garden at night.

My friends, mostly weekend gardeners
planting their six-packs of
marigolds and pansies and
veggies in the summer.
I now know who they encountered
while close to the soil.
The turtles will be pleased,
and indeed, I am too.

Collections of little critters
that counted, where critters rarely do.
Not enough credit given
to living soil.

My friends didn't choose
to ex-ter-mi-nate,
to kill everything that moves
with toxic chemicals,
running off our curbs toward the storm drain.

The popular scorched-earth approach
for sale in the aisles of your local
hardware store today,
and measureable in that fish
tonight, we may consume.

Because I have turtles
my friends collect
strange creatures too.
Why not you?

Funeral for a Turtle

Here I, a boy at fault,
standing sadly by
a funeral for a turtle.
First pet, first prize and
despite my care, soon died.

It became a riddle
that bedeviled me, how
to keep a turtle alive
would remain on my brow.

It's a mystery now,
to look in youth's direction,
for answers about death,
a narrowing recollection.

Given another chance I said
I would do it differently,
and eventually did succeed
in keeping turtles that may,
my life span exceed.

Though given another chance,
I remain today that boy at fault,
standing sadly by
a funeral for a turtle.

Myths and Folklore

Myths and folklore
of turtles abound,
turtles keep going
with hardly a sound.

Of turtles abound
myths and folklore
with hardly a sound
turtles keep going.

Turtles keep going
with hardly a sound.
myths and folklore
of turtles abound.

With hardly a sound
turtles keep going
of turtles abound
myths and folklore.

Gracie

It's time to get up pop because
it's pond-cleaning day!
It's pond-cleaning day!
Her most favorite day of the week,
and she knows that today
is pond-cleaning day,
because after feeding her I go back to sleep.

Gracie is a turtle dog.
I'm not sure if it's work or play,
but when I siphon the ponds, smelly and
green,
there'll be no keeping her away.
She's focused on the restless turtles:
restless because their water level's low.
Now she's running from one pond to the next,
panting excitedly as she goes.

I now know there are other turtle dogs –
Boykin Spaniels can sure hold their own.
Back then, I'd never heard of such a thing,
until we brought that turtle dog home,
and life began to change
in ways I could have never known.

From the Ukiah shelter at age eight,
too long in the shelter was she.
Life for Gracie had not been great,
how could this be chance we meet?
I saw her first on Craigslist,
from her eyes I knew her heart was good.
She'd been waiting for a turtle guy like me,
so it turned out, exactly, as it should.

When we walked from the shelter together
our family now complete,
Gracie did not pause, nor glance back,
and like she did it every day,
jumped into the back seat.

No Best in Show will she compete:
head too small, and graying chin,
her body a misshapen sausage
with miscellaneous conditions of the skin.
With paws too small and legs too thin
for her body, man that dog could run!
To lead you happily to the turtles,
as if she'd been shot out of a gun.

And never were eyes so brown and soft
sweetly imploring, she had no need to talk,
we could not refuse her a pet or scratch,
or to take her on her beloved walk.

Sometimes the turtles made her anxious,
and she couldn't help but bark and whine.
Now she's off to investigate the other ponds –
who knows what those turtles have in mind?
The turtles choose to ignore her,
not willing to abandon their bask.
The sun feels so good on cold turtle shells,
they fail to appreciate her task.

Gracie is a turtle dog,
and today she *earned* her keep.
Now for her reward a late afternoon nap,
the turtles set for another week.
She will always be a turtle dog.
Look, she's chasing them in her sleep.

Because I Think Of

Because I think of the earth revolving on the
back of a giant tortoise, I imagine perfection.

Because I think of water flowing swiftly,
carrying all turtles to safety, far beyond the
poacher's net, I imagine hope.

Because I think of the sun warming the black
shells of the South American giant river
turtles, who had left the river as one group,
talking, reassuring, in tones too low for an
adult human ear, I imagine community.

Because I think of the air that surrounds us,
drifting upon the surface of a pond, where a
turtle's head is breaking that surface,
breathing the same life-giving oxygen, I
imagine a shared destiny.

Because I think of a child crying at the creek's
edge, clogged with the bodies of frogs, fish and
turtles, killed by the spill of toxics that were a
byproduct of things humans throw away
without thinking, I imagine despair.

And because I think, I sometimes get angry,
then slowly in defense of perfection I begin to
shoulder my weight against a force that is
difficult to describe or stop. I then imagine I
have strength.

Conclusion

Little turtle and I
(so very young),
green (both, behind the ears),
eyes, reflecting sunlight,
the morning air clean,
our future that day
looked bright.

Little turtle and I
(now grown),
green, (both, still),
eyes restless, still searching,
anticipating each season
with a yearning.

Little turtle and I
(now old),
green (forever green),
light from eyes fading
and memory
still strong.
Memories of youth and turtles
we're both still hopeful
for tomorrow.

Index

Michael Christensen has studied turtles and tortoises for over forty years; he remains active in their care, advocacy and rescue. Michael had been writing these poems for ten years when he published the first version of this book in 2004. Because he continued to write them, and because he finally got around to getting an ISBN number, he released this updated version.